AI WITHOUT FEAR

A No-Stress Guide to Living and Working Well with AI

Hermes Hammes

AI WITHOUT FEAR

A No-Stress Guide to Living and Working Well with AI

The information in this book reflects the author's views at the time of publication. Artificial intelligence tools, platforms, and practices evolve rapidly, and some examples may change over time. This does not affect the core principles discussed.

ISBN: **978-1-971603-01-8**

First Edition

Printed in the United States of America

"The danger is not AI becoming smarter. The danger is humans becoming passive." (From the book, AI Without Fear)

"Freedom is not infinite choice. It is a meaningful choice." (From the book, AI Without Fear)

"The future does not happen to you. Millions of small, ordinary decisions shape your future". (From the book, AI Without Fear)

CONTENTS

Introduction 1

1. What AI Really Is (In Plain Human Language) ... 5

2. Why Everyone Is Afraid of AI (and Why That's Normal) 15

3. AI Is Not Here to Replace You, It Is Here to Assist You 26

4. Using AI to Work Less (Not Faster)..................... 33

5. AI at Work: Simple Uses for Everyday Jobs....... 38

6. AI at Home: Life Admin, Parenting, and Mental Space... 45

7. Creativity, Writing, and Ideas: AI as a Partner . 59

8. The Attention Trap: When AI and Technology Control You 73

9. Ethics, Truth, and Responsibility (Without Philosophy Headaches) 89

10. How to Start Using AI Without Overwhelm ... 105

11. Staying Relevant in an AI World 116

12. A Human Future With AI 132

A Final Reframing 150

What Comes Next 150

INTRODUCTION

If AI (Artificial Intelligence) feels overwhelming right now, you are not imagining it.

The pace is fast, the language is confusing, and the noise is constant. Every week, there is a new headline promising miracles or predicting disaster. One moment, AI is going to save you time; the next, it is going to take your job. Somewhere in between, you are just trying to live your life.

So let's start with something important and true: **You are not late. You are right on time.** AI did not arrive quietly. It showed up all at once, in our work, our phones, our conversations, and our fears. Moreover, unlike past technologies, it did not wait for us to be ready. That alone is enough to make anyone feel behind, anxious, or resistant.

Why AI Feels So Overwhelming Right Now

AI feels overwhelming because it touches everything at once: work, creativity, money, education, relationships, and even identity. It isn't just a new tool, it is a shift in how tasks are done, how decisions are made, and how value is measured.

On top of that, most explanations are built for engineers, investors, or early adopters. Everyone else is handed technical language, complex diagrams, and unrealistic expectations. Instead of clarity, we get confusion. Instead of help, we get pressure. And pressure is the fastest way to shut curiosity down.

The Lie That "Everyone Else Already Understands This"
One of the most damaging myths around AI is the idea that everyone else already understands it, that you somehow missed the meeting where it was all explained. You didn't. Most people are pretending.

Behind confident posts, buzzwords, and polished presentations, many are just as unsure as you are. They're experimenting, guessing, and learning in real time. The difference isn't intelligence, it is permission. Permission to start imperfectly.

This book exists to give you that permission.

What This Book Will *Not* Do

Let's be clear about what you won't find here:

- No Coding.
- No technical jargon.
- No hype, fear-mongering, or exaggerated promises.
- No pressure to "keep up" or optimize your entire life.

This isn't a book about becoming a tech expert. It is not about mastering tools for the sake of mastery. Moreover, it's definitely not about turning your life into a productivity machine.

What This Book *Will* Do

This book is about using AI to live better. It will help you understand AI in plain, human language, enough to feel confident, not overwhelmed. You will see how AI can reduce mental load, simplify work, support creativity, and give you back time and energy. You will also learn when *not* to use AI, how to set boundaries, and why human judgment matters more than ever. Most importantly, this book treats AI as what it really is: **A tool, not a replacement for your thinking, your values, or your life.**

You don't need to run faster.

You don't need to know everything.

You don't need to become someone else.

You just need a calmer, clearer way forward. And that's exactly where we'll begin.

1

WHAT AI REALLY IS (IN PLAIN HUMAN LANGUAGE)

Reflection Question:

What part of AI scares or confuses me most, and if I explained AI to a close friend in my own words, what would I realize I already understand?

AI (Artificial Intelligence) is already woven into everyday life, often in ways we don't even notice.

When you say, *"Alexa, play some music,"* and it understands you. When you scroll on your phone and the next video feels almost too perfectly matched to the one you just watched. When you ask Siri a question and it answers naturally, without you thinking about how it works. That is AI.

For many people, this technology didn't arrive with an announcement. It crept in quietly, helping, suggesting, and responding, until one day it felt like it was everywhere at once. That's where the pressure began. The confusion. The quiet fear of falling behind something you did not even realize you were already using.

If AI is already part of your daily life, the real question is not whether you are "using" it. The question is whether you understand what it is, and what it isn't. Much of the fear around AI comes from treating it as something mysterious, powerful, or almost human. However, once we slow down and strip away the buzzwords, AI becomes easier to understand. Not as a mind. Not as a replacement for people, but as a tool. A tool that follows patterns, processes information, and supports decisions without ever making them for us. Moreover, that understanding is where fear begins to lose its grip.

AI has been explained so many times that it has almost disappeared behind its own explanations. Charts, acronyms, technical terms, predictions, warnings, and promises, most of them aimed at people who already live comfortably in the world of technology. For everyone else, the result isn't clarity, but distance. AI becomes something *other*, something happening to us instead of for us.

So let's reset.

If you were explaining AI to a friend, not a conference audience, not a LinkedIn crowd, not a tech team, but a real person sitting across from you, and you would not start with algorithms or data models. You would say something like this: **AI is a tool that helps humans handle information and tasks more easily.** That's the foundation. Everything else is decoration.

AI Without the Drama

AI is not a creature. It is not alive. It does not wake up with intentions. It does not want anything.

AI is software designed to recognize patterns and produce outputs based on what it has seen before. When it appears intelligent, what you are really seeing is speed and scale, not understanding. Think of it like this: A calculator does not understand math. It processes numbers extremely fast. AI does the same thing, but with language, images, sound, and data.

A Simple, Everyday Example

Imagine you are overwhelmed at work. Your inbox is full. You are tired. You need to respond to a long email, explaining a situation calmly and clearly. You ask AI to draft a response. It gives you something usable, not perfect, but structured.

You then:

- Adjust the tone.
- Remove what doesn't feel right.
- Add context only you know.

AI didn't communicate. **You did.** AI reduced friction. **You made meaning.** That distinction matters more than most people realize.

Why AI Feels Smarter Than It Is

AI feels impressive because it compresses effort. What once took hours now takes minutes. What once required searching, organizing, summarizing, and rewriting can happen in a single interaction. That compression creates the *illusion* of intelligence. However, intelligence isn´t output. Intelligence includes:

- Judgment.
- Values.
- Context.
- Consequences.
- Responsibility.

AI has none of those.

What AI Can Do Well

AI excels at tasks that are:

- Repetitive.

- Structured.
- Information-heavy.

For example:

- Summarizing long documents.
- Drafting first versions of text.
- Organizing ideas.
- Translating languages.
- Creating outlines.
- Generating options.

Real-life example: AI save time

A small business owner spends Sunday nights writing invoices, emails, and social posts. After using AI to draft these materials, Sunday night becomes family time again. AI didn't build the business. AI did not define the brand voice. AI removed unnecessary labor. That's its role.

What AI Cannot Do (And Never Should)

AI cannot:

- Care whether something is right or wrong.
- Understand emotional weight.
- Take moral responsibility.

- Decide what truly matters.
- Replace lived experience.

It does not know grief.

It does not know fear.

It does not know love.

When AI "sounds empathetic," it is copying language patterns, not feeling anything. That's why human oversight is not optional.

AI Is Not Thinking - And That Is a Good Thing

Human thinking is messy. We contradict ourselves. We feel before we reason. We hesitate. We reflect. We change our minds. These qualities are not flaws, they are depth. AI does none of this.

For AI there is no doubt. It does not pause. It does not wonder. It predicts what comes next based on probability. Speed without reflection is not wisdom.

The GPS Lesson (Why Tools Can Mislead)

GPS is one of the best analogies for AI.

GPS:

- Knows the fastest route.
- Doesn’t know road closures.
- Doesn’t know unsafe neighborhoods.
- Doesn’t know your priorities.

When people blindly follow GPS into rivers, deserts, or dead ends, the problem is not the tool; the problem is we abandoned judgment. AI works the same way. It can guide, but It cannot decide.

Tools Assist - Humans Decide

This line is non-negotiable.

AI can:

- Suggest candidates.
- Analyze patterns.
- Generate recommendations.

However, humans must:

- Hire.
- Diagnose.

- Judge.
- Lead.
- Care.

When decisions move fully to tools, responsibility disappears, and that's where harm begins.

Why This Understanding Changes Everything

When you understand AI as a tool, fear loosens its grip.

You stop asking:

- "Will AI replace me?"

And you start asking:

- "What should I delegate?"
- "Where does my judgment matter most?"
- "How do I want to use my time?"

AI does not shrink your humanity. It reveals where it matters most.

The Calm Truth

AI is not here to replace your thinking. It is here to remove unnecessary weight. The danger is not AI becoming smarter. The danger is humans becoming passive. And that is always a choice.

Pause for a Moment before Moving On

AI can feel abstract, intimidating, or distant, but clarity begins when we slow down and name what we actually understand.

Reflect:

- In your own words, how would you explain AI to someone you trust?
- What assumptions about AI have softened or changed after this chapter?
- What feels clearer now than it did before?

Gentle Reminder:

You don't need technical mastery to engage thoughtfully with technology. Understanding starts with language you can live with. You don't need to be an expert to belong in the future. You just need a thoughtful way to engage with it.

2

WHY EVERYONE IS AFRAID OF AI (AND WHY THAT'S NORMAL)

Reflection Question:

When I think about AI, what am I actually afraid of losing, and how much of that fear is about technology versus my sense of security, identity, or control?

Fear is often treated as something to overcome quickly, something to silence or dismiss. However, fear has a purpose. It shows up when something meaningful is shifting, when the ground beneath our habits begins to move. In that sense, fear isn't a flaw. Fear is a signal.

The fear surrounding AI is not irrational. It is a deeply human response to rapid change arriving all at once, without clear rules, shared language, or enough time to adapt gradually.

If AI makes you uneasy, you are not weak. You are paying attention. To move forward calmly, we must first understand what we are actually afraid of.

Fear Has a Pattern

Most major shifts follow a familiar emotional sequence:

- Confusion.
- Fear.
- Resistance.
- Adaptation.
- Integration.

We tend to remember history from the point of integration, when a technology feels obvious and necessary, but people living through the early stages rarely feel calm. They feel threatened, disoriented, and uncertain.

AI is currently sitting between fear and resistance. That is exactly where it is supposed to be.

The Fear of Job Loss

This fear is immediate, practical, and deeply personal. Work isn't just how we survive financially. It is how we measure contribution, competence, and value. When people hear that AI can write, analyze, design, summarize, or plan, the unspoken question isn't technical. It is existential: **If a machine can do part of what I do, am I still needed?**

This fear shows up differently depending on life stage:

- A young professional worries their career path is already outdated.
- A mid-career worker fears becoming replaceable after years of experience.
- A business owner worries competitors will adopt faster tools.

These fears aren't imaginary. They are rooted in real economic pressure. However, here is what´s often left out: **AI replaces tasks, not meaning.** It shifts work; it does not erase the need for humans.

Real-life example: AI transform work

A legal assistant hears that AI can draft contracts and summarize case files. Panic sets in, but when AI is introduced, the role changes. Less time spent formatting and searching. More time is spent verifying, communicating with clients, managing nuance, and spotting risk.

The job becomes more human, not less. Work has never disappeared. It has transformed.

The Fear of Irrelevance

This fear cuts deeper than job loss. It sounds like:

- “I’m too late to learn this”
- “This world isn’t built for people like me”
- “I don’t speak this language”

Irrelevance is not about ability. It’s about belonging.

AI often arrives wrapped in jargon and unspoken expectations. People who do not immediately understand feel excluded from the future, and exclusion hurts more than ignorance.

Real-life example: Being invisible

A highly experienced manager watches younger colleagues use AI fluently. No one explains. No one invites questions. The fear isn't stupidity. It's invisibility. However, here is the truth: **AI favors clarity, not youth.** It rewards experience, judgment, and context. Relevance is not about speed. It's about discernment.

The Fear of Being Controlled by Technology

This fear often shows up as cynicism or resistance:

- "I don't trust algorithms."
- "I don't want machines deciding my life."
- "I already feel controlled by screens."

This fear is earned.

Technology has already shaped attention, behavior, and identity. Social media algorithms influence what we see and how we compare ourselves. So when AI enters the conversation, it can feel like the next step toward losing autonomy.

Real-life example: Technology takes over

Isabella wakes up and checks their phone before checking in with themselves. Notifications shape mood. Time disappears. The fear is not AI. It is the loss of agency. Here is the distinction that matters: Technology does not take control. **Unconscious use hands it over.**

AI does not force decisions. Humans abdicate responsibility when they stop choosing intentionally. The danger is not AI becoming powerful. The danger is humans becoming passive.

Why Fear Always Appears During Big Shifts

History is consistent on this point:

- When the printing press spread, leaders feared loss of authority.
- When electricity arrived, people feared disruption of natural rhythms.

- When radio and television emerged, families feared moral decay.
- When computers entered offices, workers feared mass unemployment.
- When the internet arrived, people feared the end of real connection.

Each fear contained some truth and much exaggeration.

Fear often appears when:

- Systems change faster than culture.
- Tools spread faster than understanding.
- Power feels redistributed.

Fear is the nervous system responding to uncertainty. The mistake is treating fear as prophecy instead as information.

Fear vs. Reality

Fear says: "This will take something from me." Reality says: "This will change how value is expressed." AI does not erase human strengths. It reveals which strengths matter most:

- Empathy.

- Judgment.
- Creativity.
- Ethics.
- Context.
- Responsibility.

These cannot be automated.

The Psychological Weight of Speed

AI advances in weeks, not decades. There is no shared pause, no collective breath. When change arrives faster than identity can adjust, anxiety follows. This isn't personal failure, it is biological reality.

The answer isn't to move faster. The answer is to slow your **relationship** with change, even if change itself doesn't slow down.

Fear Is a Signal, Not a Verdict

Fear tells you:

- Something matters.
- Something is uncertain.

- Something needs attention.

It does **not** tell you:

- To stop.
- To hide.
- To surrender judgment.

The healthiest response to AI is neither blind optimism nor total rejection. It is **a caution**.

Ask:

- What helps me here?
- What replaces effort but not meaning?
- What do I want to remain human?

From Fear to Agency

When people understand AI more clearly, fear begins to shift. They stop asking:

- "Will I be replaced?"

And they start asking:

- "What do I want help with?"
- "Where does my judgment matter most?"
- "How do I want to spend my limited time?"

Agency returns when fear is named, not denied.

The Calm Truth

AI is not the first disruption, and it won't be the last. What matters is not whether AI advances. What matters is whether humans remain awake, responsible, and intentional.

Fear will show up. That doesn't mean it gets to lead. You do not need to eliminate fear to move forward. You need to walk with it, instead of letting it decide for you. In addition, when you do, fear becomes something else entirely: not an obstacle, but a guide pointing toward what matters most.

Pause for a Moment before Moving On

Fear is not a failure. It's a signal. This chapter is not asking you to eliminate fear, but to listen to it honestly.

Reflect:

- What specific fears came up for you while reading this chapter?
- What do those fears say about what you value or want to protect?

- Which fears feel grounded, and which feel inherited from headlines or noise?

Gentle Reminder:

Fear often appears at moments of transition. Feeling it does not mean you are behind, it means you are paying attention

3

AI IS NOT HERE TO REPLACE YOU, IT IS HERE TO ASSIST YOU

Reflection Question:

If AI could quietly assist me without taking away what makes me human, which parts of my work or life would I want help with, and which parts would I never hand over?

Replacement is the word that causes panic. When people hear “AI,” their minds don’t jump to help or support. They jump to displacement, to being pushed aside, to becoming unnecessary. That reaction makes sense, because for a long time value has been tied to output: how much you produce, how fast you respond, how efficiently you perform.

So when a tool appears that can produce faster than a human can, fear feels logical. That fear rests on a false assumption: **Are humans valuable primarily because of speed and output?** They are not.

AI is not a competitor for your humanity. It's a mirror, forcing a deeper question: **What part of my work actually requires me?**

The Assistant Mindset

The healthiest way to approach AI is with an *assistant mindset*. An assistant does not lead your life. An assistant does not set goals. An assistant does not decide what matters. An assistant handles support tasks so you can focus on what only you can do.

A good assistant:

- Prepares information.
- Organizes details.
- Reduces friction.
- Anticipates needs.

A good assistant does **not** replace judgment. It amplifies it. AI works best in exactly this role; it amplifies what is strictly human.

The problem begins when people adopt a *replacement mindset* instead, when the tool starts to feel like an authority rather than support. That's when anxiety grows and responsibility slips away.

A Real-Life Example: Using AI as an Assistant

A project manager in a mid-sized company feels overloaded.

Before AI:

- Emails consume half the day.
- Meeting notes pile up.
- Reports take hours.
- Mental energy is depleted.

After using AI as an assistant:

- Emails start as drafts, and become human messages.
- Meetings turn into clear action lists.
- Reports begin with structure.

- Mental energy is preserved for leadership.

The job does not disappear. It becomes more human. Less typing. More thinking. Less busywork. More judgment.

AI as a Time-Saver, Not a Boss

One of the most damaging myths is that AI should "optimize everything." Optimization can hide a dangerous assumption: "faster is always better". It isn't. Speed without intention leads to burnout, not freedom. AI should not tell you how to work. It should give you back time to decide *why* and *how* you work.

Why Human Judgment Still Matters

AI can generate possible answers. Humans must choose the right one. Judgment involves:

- Context.
- Consequences.
- Nuance.
- Responsibility.

AI cannot do these things because it does not live with outcomes. When judgment is removed, decisions become technically efficient but humanly flawed.

AI without judgment doesn't lead to progress. It leads to detachment.

Empathy: The Skill AI Cannot Fake

AI can imitate empathetic language. It cannot be empathetic. Empathy requires presence, emotional awareness, lived experience, and moral responsibility.

In healthcare, education, leadership, parenting, counseling, and caregiving, empathy is not optional, it is foundational. AI can support these roles with information. It cannot replace the human connection at their core.

Creativity Is More Than Output

AI can generate art and writing, but creativity is more than creation. Creativity is expression, born from memory, emotion, struggle, desire, and meaning. AI recombines. Humans transform experience into insight.

The Jobs AI Cannot Truly Replace

AI may change how jobs function, but entire categories of work cannot be fully replaced because they depend on presence, trust, and accountability:

- Care-based roles.
- Leadership roles.
- Creative roles.
- Judgment-based roles.
- Relationship-driven roles.

AI can assist these roles. It cannot embody them.

The Calm Truth

AI is not here to take your place. It is here to take your burden. It handles the repetitive, draining, mechanical tasks, so you can focus on what requires awareness, care, and intention.

The future does not belong to those who compete with machines. It belongs to those who collaborate wisely with them. You don't need to become faster than AI. You need to become more human where it matters.

Pause for a Moment before Moving On

This chapter reframes AI not as a competitor, but as support. That distinction matters more than it may seem.

Reflect:

- Where in your life do you feel overextended or stretched thin?
- What kind of support would actually help you, not replace you?
- What parts of your work or identity feel deeply human and non-negotiable?

Gentle Reminder:

You are not valuable because of how much you produce. Your judgment, care, and presence cannot be automated.

4

USING AI TO WORK LESS (NOT FASTER)

Reflection Question:

Which daily tasks drain my energy without adding meaning, and what would change if they took up less mental space?

For years, we were taught the goal of work was productivity:

- Produce more.
- Respond faster.
- Optimize every minute.
- Fill every gap.

Technology promised to help, and in many ways, it did. Yet despite all this efficiency, people are more exhausted than ever.

This chapter is not about becoming faster. It is about becoming lighter. AI's real value is relief us from cognitive overload, decision fatigue, and the pressure to always be "on."

Productivity vs. Exhaustion

Productivity is measured in output. Exhaustion is measured in what it costs you. Many people are productive and exhausted at the same time. That's the trap.

Real-life example: Mental Load

Taylor answers emails quickly, attends nonstop meetings, and completes tasks, yet still ends the day drained and foggy. Nothing is technically "wrong," but something is deeply unsustainable. The problem is not laziness. It is a mental load.

What Mental Load Really Is

Mental load is the invisible work of holding everything in your head:

- What needs to be done?
- When is it due?
- Who is waiting?
- What is already decided?
- What still needs deciding?

Mental load doesn't clock out. It follows you home. AI's most underrated benefit is not speed. It is offloading.

Delegating Boring, Repetitive Tasks

Every job includes tasks that quietly drain energy:

- Rewriting similar emails.
- Formatting documents.
- Summarizing notes.
- Organizing scattered thoughts.

AI can take the first pass, so you can bring judgment, clarity, and voice.

Creating Breathing Room

The breathing room is not empty. It is unpressured time-space without urgency. This is where insight returns, creativity comes back, and calm replaces reaction. AI should create space, not fill it.

The Calm Truth

AI's greatest gift is not speed. It is space.

- Space to think.
- Space to rest.
- Space to decide intentionally.

When you use AI to work less, not faster, you protect your humanity. That´s the most valuable work you will ever do.

Pause for a Moment before Moving On

Working less does not mean caring less. It means caring wisely. This chapter invites you to rethink effort, exhaustion, and productivity.

Reflect:

- Which parts of your day feel draining without being meaningful?
- What would it feel like to end the day with more energy than you started with?
- Where could relief and change your experience of work?

Gentle Reminder:

Rest is not a reward for productivity. It's a requirement for being human.

5

AI AT WORK: SIMPLE USES FOR EVERYDAY JOBS

Reflection Question:

In my own work, what task feels like it takes more effort than it should, and what would it feel like to have quiet support instead of doing everything alone?

When people talk about AI at work, the conversation often swings to extremes. Either AI will automate everything, or it will destroy jobs. Both miss what is actually happening.

For most people, AI is arriving quietly, solving small problems, relieving pressure, and supporting daily tasks. This chapter is not about tools. It is about real-life use.

AI at Work Is Not About Becoming Someone Else

AI does **not** require you to:

- Become technical.
- Change your personality.
- Abandon your profession.
- Compete with machines.

AI adapts to how you already work, if you let it.

The most effective uses are not flashy or dramatic. They simply make work feel easier and more humane.

Office Workers: Reducing Noise, Not Responsibility

Office work is demanding because of volume, emails, meetings, documents, updates, and follow-ups. AI helps by organizing the background:

- Turning summaries into action items.
- Drafting routine messages.
- Converting notes into clear outlines.

The work does not disappear. It becomes less draining.

Entrepreneurs and Small Businesses: Wearing Fewer Hats

Entrepreneurs are overloaded by roles. AI can reduce friction by:

- Drafting marketing text.
- Organizing schedules.
- Summarizing customer feedback.

It does not replace vision. It protects it.

Freelancers and Creators: Protecting Energy and Voice

Creators face constant pressure to produce. AI can:

- Start the draft.
- Organize research.
- Break blank-page paralysis.

AI handles the start. The human handles the soul.

Teachers, Coaches, and Caregivers: Making Space for Presence

In teaching and care work, time is often swallowed by planning and administration. AI can:

- Draft lesson or session outlines.
- Generate discussion prompts.
- Adapt materials for different needs.

AI does not replace teaching, but protect presence.

Real-life example: Overwhelm by planning

Mary is a middle-school teacher who loves being in the classroom, but dreads what happens after the students leave.

Every afternoon, her energy was drained not by teaching, but by everything around it: lesson planning, adapting materials for different learning levels, writing discussion questions, and preparing notes she rarely had time to revisit. By the time she finished, she felt depleted and less present the next day.

She began using AI in a very limited, intentional way. Once a week, she asked AI to:

- Draft a rough lesson outline based on the topic she was already teaching.
- Suggest discussion questions to spark curiosity.
- Adapt the same lesson for students who needed simpler explanations or extra support.

She didn't use those outputs as final material. She adjusted them. Removed what didn't fit. Added her own voice and experience.

What changed wasn't her teaching style. What changed was her energy. Instead of spending hours preparing, she spent more time noticing students who were struggling, responding more patiently to questions, and being fully present during discussions. The classroom felt calmer, not because it was optimized, but because she was less exhausted.

AI didn't replace her role as a teacher. It quietly protected what mattered most: her attention, her care, and her presence.

The Calm Truth

AI does not redefine work. It reveals where work has become inhuman.

Used wisely, AI does not make you replaceable. It makes you available to think, connect, lead, and care. In a world overwhelmed by speed and noise, that availability isn't a weakness. It's power.

Pause for a Moment before Moving On

AI doesn't need to transform your career to be useful. Sometimes, small shifts are enough.

Reflect:

- What recurring task at work feels heavier than it should?
- Where do you spend time managing instead of thinking?
- How would your workday change if one small friction point disappeared?

Gentle Reminder:

You don't need to optimize everything. Improving even one experience is already progress.

6

AI AT HOME: LIFE ADMIN, PARENTING, AND MENTAL SPACE

Reflection Question:

What invisible mental load am I carrying at home right now, and how would my relationships or presence change if my mind had a little more space?

For many people, the most exhausting part of life isn't work. It is everything else. It's the endless background noise of daily living, the appointments, the groceries, the school emails, the meals that need planning, the birthdays you forgot, the decisions that pile up quietly until your brain feels like a browser with too many tabs open.

AI doesn't enter our homes as a robot vacuum with glowing eyes or a talking assistant perched on the counter. It enters invisibly, through lists, reminders, suggestions, drafts, plans, and quiet help behind the scenes.

When used well, AI at home isn't about automation for its own sake. It is about mental relief. It is about reclaiming attention, presence, and energy, especially for the people and moments that matter most.

This chapter is not about turning your household into a machine. It's about making space to be human inside it.

The Hidden Weight of Life Administration

Life admin is the work no one trained us for, no one pays us for, and no one escapes. It includes:

- Managing schedules and calendars.
- Coordinating family logistics.
- Tracking bills, renewals, and subscriptions,
- Planning meals and grocery lists.
- Remembering school requirements and deadlines.

- Making small decisions all day long.

Individually, these tasks seem trivial. Collectively, they are overwhelming.

Psychologists call this **cognitive load**, the total amount of mental effort being used at any given time. When cognitive load is high, even small decisions feel heavy. You forget things you "should" remember. You feel tired before the day has even started.

AI's greatest contribution at home isn't speed. It is a relief from invisible labor.

Managing Household Tasks Without Becoming a Project Manager

Most households already run like small companies, without the tools. Think about it:

- There are recurring tasks.
- Shared responsibilities.
- Deadlines.
- Budgets.
- Coordination between multiple people.

Instead of systems, many families rely on memory, sticky notes, text messages, and stress.

Real-Life Example: The Overloaded Parent

Take Maria, a working parent with two kids. Her problem wasn't that she didn't know what needed to be done. She knew everything:

- School lunches.
- Soccer practices.
- Doctor appointments.
- Laundry schedules.
- Grocery runs.
- Birthday gifts.
- Homework deadlines.

Her problem was that **everything lived in her head**.

When Maria began using AI simply as a thinking partner, not as a boss, something shifted. She started by asking: "Help me list all recurring household tasks in a normal week."

The AI returned a structured list, daily, weekly, monthly. Seeing it laid out visually did two things:

1. It validated her exhaustion. “No wonder I’m tired”.
2. It made delegation and simplification possible.

Then she asked:

“Which of these tasks can be simplified, batched, or automated?”

The result wasn’t perfect. It was clear. AI didn’t clean the house. It didn’t parent the kids. It didn’t take responsibility away. It helped her stop carrying everything alone.

Planning Meals Without Burning Out

Few daily tasks drain energy like answering the question: “What are we eating tonight?” Multiply that by seven days a week, plus dietary needs, budgets, preferences, and time constraints, and meal planning becomes a mental loaded.

How AI Helps Without Taking Control

Used well, AI doesn't dictate meals. It removes the blank page. You can ask:

- "Create a 5-day dinner plan for a busy family with 30-minute meals."
- "Suggest meals using chicken, rice, and vegetables I already have."
- "Help me plan a week of meals under $120."

Real-Life Example: Decision Fatigue at Dinner

James and his partner realized their biggest daily arguments happened around food, not because of the food itself, but because they were both exhausted.

They started using AI once a week on Sunday night. Together, they:

- Set basic preferences.
- Agreed on a budget.
- Asked AI to generate options, **not commands.**

Then they chose.

This small shift removed dozens of micro-decisions during the week. Dinner became execution, not debate. The key difference stayed simple: **AI suggested. Humans decided.**

Planning Trips, Routines, and the "In-Between" Moments

Life isn't just big events. It's the in-between moments that wear us down:

- Morning routines.
- After-school chaos.
- Packing for trips.
- Coordinating schedules.

AI excels at structure, sequencing, and reminders, things humans are bad at when tired.

Family Routines That Actually Work

Routines fail when they are too rigid or unrealistic. AI can help design routines that fit real life:

- "Help me design a calm morning routine for a family with two kids under 10."

- "Create a bedtime routine that reduces screen time without constant conflict."

These aren't prescriptions. They're drafts.

Parents who succeed with AI don't follow plans blindly. They adjust, discard, and adapt. The win isn't compliance. It reduces friction.

Helping Kids Learn Without Replacing Thinking

This is where many parents feel conflicted. On one hand, AI can help explain concepts, generate practice problems, and support learning. On the other hand, parent wants their child to think. The solution lies in **how** AI is used, not whether it is used. **AI as a Tutor, Not an Answer Machine.**

The healthiest approach positions AI as:

- A guide
- A clarifier
- A question-asker

Not a shortcut.

Examples:

- "Explain this math concept in a simpler way."
- "Give hints, not the answer."
- "Ask my child questions that help them think this through."

Real-Life Example: Homework without Tears

Lena's son struggled with math. Homework time turned into nightly frustration, on both sides.

Instead of asking AI to solve problems, Lena asked: "Explain this concept in three different ways, using examples a 10-year-old would understand."

They read the explanation together. The AI didn't replace her role as a parent. It supported it. Even more important, Lena set a rule: **AI could help explain, but it could never submit answers.**

The result was real:

- Less conflict.
- More understanding.

- A child who still had to think.

And homework was done.

Teaching Kids How to Use AI Responsibly

Children will grow up with AI whether we like it or not. The goal isn't avoidance. It's literacy.

Parents who integrate AI thoughtfully can teach:

- How to ask better questions.
- How to evaluate answers.
- How to think critically instead of copying blindly.

This mirrors how calculators were introduced, not as thinking replacements, but as tools. AI becomes a mirror for thinking, not a substitute for it.

Reducing Decision Fatigue: The Quiet Superpower

Decision fatigue is real. Every decision, even small ones, uses energy. By the end of the day, your ability to choose wisely drops dramatically. AI can't live your life for you.

However, it can reduce the number of decisions you have to make.

Examples:

- Clothing suggestions for the week based on weather.
- Default grocery lists.
- Pre-decided routines.
- Reusable templates for common tasks.

The goal isn't to remove choice. It is to remove unnecessary choices.

Mental Space Is the Real Productivity

The biggest benefit people report after using AI at home isn't efficiency. It's mental quiet:

- Fewer forgotten tasks.
- Less background anxiety.
- More presence with family.
- Less guilt about "falling behind".

AI doesn't create more time. It creates less noise inside the time you already have.

Boundaries Matter More Than Tools

Without boundaries, AI becomes just another source of overwhelm.

Healthy households set rules:

- No AI during family conversations.
- No replacing thinking with answers.
- No constant optimization.

AI should support life, not turn it into a performance.

The Human Remains at the Center

Your home doesn't need to be smarter. It needs to be calmer.

AI works best when it:

- Handles structure.
- Holds information.
- Reduces mental clutter.

And steps back when it comes to:

- Love.
- Presence.
- Judgment.
- Values.

Those things cannot be automated.

You Are Not Failing, You're Carrying Too Much

If home life feels overwhelming, it is not because you are bad at it. It is because modern life asks individuals to manage systems that used to be shared across communities.

AI doesn't fix that. However it can help you carry the load more gently. That's enough.

Pause for a Moment before Moving On

Much of life's hardest work is invisible. This chapter names that truth.

Reflect:

- What mental load are you carrying that no one else sees?
- How does that load affect your presence with the people you love?
- What would mental space, not perfection, look like at home?

Gentle Reminder:

Feeling overwhelmed doesn't mean you're failing. It often means you are carrying more than one person was meant to hold.

7

CREATIVITY, WRITING, AND IDEAS: AI AS A PARTNER

Reflection Question:

If I didn't have to be perfect, impressive, or original right away, what would I want to create, and how might a judgment-free partner help me begin?

Creativity has always been surrounded by fear. Fear of running out of ideas. Fear of not being original enough. Fear of not being good enough. Fear of the blank page staring back, silently asking for something you are not sure you have.

Therefore, when AI entered the creative space, it did not just trigger excitement, it triggered panic. Writers worried they would be replaced. Artists worried originality would disappear. Creators worried their voice would be diluted, copied, or automated out of existence.

However, beneath the noise and headlines, something quieter began to happen. People who had ideas but no confidence began writing. I am one of them. People who had stories but no structure began shaping them. People who felt blocked, stuck, or behind began creating again. Not because AI was more creative than humans were, but because it removed the fear that was stopping them from starting.

This chapter is not about shortcuts or cheating. It is about what happens when pressure is removed, when the blank page stops feeling like a test, and creativity becomes a conversation instead of a performance.

The Myth That AI Kills Creativity

The belief that tools destroy creativity is not new. When photography emerged, painters' feared art was over. When digital music tools appeared, musicians' feared authenticity was lost. When word processors replaced typewriters, writers feared the craft would be cheapened. None of that happened. What changed was who could participate.

Creativity did not shrink. It expanded. AI follows the same pattern. Creativity does not come from difficulty. It comes from expression, curiosity, and connection. Difficulty doesn't create creativity, it gate keeps it. AI doesn't kill creativity. It removes unnecessary friction so creativity can surface.

Creativity Isn't about Talent, It Is about Safety

Most people don't struggle with ideas. They struggle with permission. Permission to be bad at first. Permission to experiment. Permission to explore without being judged.

The reason so many people abandon creative projects isn't lack of ability, it is because of fear. Fear of wasting time. Fear of sounding foolish. Fear of starting something they won't finish.

AI creates a private, judgment-free space. No audience. No critique. No embarrassment. Just a place to try. In addition, that safety changes everything.

Brainstorming Without Pressure

Traditional brainstorming often fails because it is public and performative.

In meetings, only the loudest voices speak. In classrooms, only the confident raise their hands. In creative rooms, people censor themselves before ideas even form.

AI removes the social risk. You can offer half-formed thoughts. You can ask "dumb" questions. You can explore ideas that feel strange, personal, or unfinished. Nothing is wasted. Nothing is judged.

Real-Life Example: The Entrepreneur Who Felt "Out of Ideas"

Daniel had been running a small business for years. When it came time to refresh his brand and messaging, he froze. "I'm just not creative," he told himself.

However, when he started using AI to brainstorm, casually and imperfectly, something shifted. He didn't ask for brilliance. He asked for:

- "Ten rough ideas, no matter how bad".
- "Unexpected ways to talk about my work".
- "How to explain this to someone who's never heard of it".

What came back wasn't perfect, but it sparked recognition. "That's wrong, but close." "That's not it, but it reminds me of something." "That's interesting, I hadn't thought of that." Creativity returned not as lightning, but as momentum.

AI as a Thought Mirror

One of AI's most underrated creative benefits is reflection. When you articulate an idea, even poorly, and see it mirrored back, your brain responds.

You notice what feels off. You recognize what feels true. You refine, clarify, and adjust. This is how thinking works: not in isolation, but in response.

AI doesn't replace the thinker. It gives the thinker something to respond to.

Writing Without Blank-Page Anxiety

The blank page is intimidating because it demands perfection upfront. It doesn't say, "Try something." It says, "Produce."

AI changes that dynamic. Instead of staring at emptiness, you start with *something*: a draft, an outline, a messy paragraph. From there, writing becomes editing, and editing is far less frightening than creation from nothing.

Real-Life Example: The Aspiring Writer Who Never Finished

Sofia had notebooks full of ideas but had never finished a piece of writing. She didn't lack imagination. She lacked momentum. Using AI, she started differently.

She would say:

- "Help me write a terrible first paragraph".
- "Create an outline I can disagree with".
- "Start this story badly so I can fix it".

For the first time, the pressure lifted.

She wasn't trying to be great. She was trying to respond. Six months later, she finished her first short story. Not because AI wrote it, but because AI helped her start without fear.

Drafting Is Not the Same as Creating

One of the biggest misunderstandings about AI-assisted writing is confusing drafting with authorship. Drafting is mechanical. Authorship is intentional.

AI can generate words. Humans decide the meaning. A raw AI draft is not a finished thought. It is clay. The creative act lives in:

- What you remove.
- What you emphasize.
- What you reshape.
- What you refuse to say.

Creativity lives in judgment, not output.

Keeping Your Voice Human

A common fear is: "What if my writing starts to sound generic?" That fear is valid. AI has patterns. Your voice has history. The key isn't letting AI lead the voice, it's letting it support the process.

Practical ways to protect your voice:

- Rewrite drafts in your own words.
- Ask AI for structure, not tone.
- Use AI to generate options, not finals.
- Read aloud and adjust rhythm.

Your voice isn't just vocabulary. It is cadence, perspective, and emotional honesty. AI doesn't have those unless you give them.

AI as a Creative Partner, Not a Ghostwriter

Partnership implies collaboration. You bring:

- Lived experience.
- Emotion.
- Intuition.
- Values.
- Taste.

AI brings:

- Speed.
- Options.
- Structure.
- Variation.

When these roles stay clear, creativity flourishes. When they blur, work becomes hollow. The most satisfying AI-assisted creative work still feels personal, because it is.

Creativity for "Non-Creative" People (Who Actually Are Creative)

Many adults stopped creating because someone once told them they were not good at it. AI reopens that door quietly.

You don't need to identify as:

- A writer.
- An artist.
- A creative.

You just need something you want to express. Emails become clearer. Presentations become more engaging. Personal reflections become more honest. Creativity doesn't have to be public to be real.

Ideas Are Cheap, Insight Is Rare

AI can generate endless ideas. That doesn't make ideas worthless. It makes insight more valuable.

Insight comes from:

- Knowing which ideas matter.

- Understanding your audience.
- Choosing what aligns with your values.
- Saying no to most options.

AI expands possibilities. Humans choose direction.

Creativity Is Not a Race

One hidden danger of AI is comparison. When content floods the world faster than ever, it is easy to feel behind. However, creativity isn't about volume. It is about resonance.

The work that lasts isn't the fastest. It is the most honest. AI doesn't change that.

When to Step Away From AI

Healthy creative practice includes restraint. There are moments when AI helps, and moments when it distracts. You might step away when:

- You need emotional clarity.
- You are processing something personal.
- You are exploring intuition, not structure.

- You want silence, not suggestions.

AI should never drown out your inner voice. If it does, pause.

The Return of Play

Perhaps the greatest gift AI offers creativity is playfulness. When outcomes matter less, exploration returns.

You try different styles. You experiment with tone. You explore ideas you would never share. Playfulness is where creativity thrives. AI gives permission to play again.

You Are Still the Source

No matter how advanced tools become, they don't originate meaning. They rearrange patterns. They echo existing ideas. They respond to prompts.

Meaning comes from:

- What you have lived.

- What you care about.
- What you choose to express.

AI can amplify creativity. It cannot replace humanity.

Creativity Was Never the Problem

If you have felt blocked, stuck, or uninspired, creativity didn't leave you. It was buried under pressure, judgment, and exhaustion.

AI doesn't create for you. It clears space so you can create again. Creativity thrives where pressure disappears.

Before Moving on, Take a Moment

This chapter isn't about output. It is about permission.

Reflect:

- When was the last time you created without worrying about quality or outcome?
- What ideas have you avoided because starting felt too heavy?

- How might curiosity change your relationship with creativity?

Gentle Reminder:

Creativity isn't something you lose. It is something that waits for safety.

8

THE ATTENTION TRAP: WHEN AI AND TECHNOLOGY CONTROL YOU

Reflection Question:

When I finish using technology, do I usually feel clearer or more scattered, and what boundaries would help me protect my attention instead of constantly spending it?

Technology didn't take over our lives overnight. It happened in fragments. A notification here. A convenience there. A tool that saves time. Another that promised efficiency. Another that promised inspiration. None of it felt dangerous in isolation.

Yet many people wake up each day already tired, before anything has even happened. Not because life is harder than before, but because **attention has become the most contested resource of our time**. AI didn't invent this problem. However, AI accelerated it.

This chapter isn't about rejecting technology. It is about recognizing when technology stops serving you, and starts shaping you.

The Illusion of Control

Most people believe they are in control of their technology. They choose their apps and tools. They choose when to engage. However, control is not about choice. It is about awareness.

The attention economy does not require you to lose control completely, only to lose it gradually. That's what makes it so effective.

Overuse vs. Mindful Use

The difference between helpful technology and harmful technology is not the tool. It is the relationship. Overuse is not defined by hours alone. It's defined by compulsion, fragmentation, and depletion.

Mindful use feels different in the body:

- You feel calmer afterward, not emptier.
- You feel clearer, not scattered.
- You feel supported, not pressured.

Real-Life Example: The Constant Optimizer

Ethan loved tools. He followed productivity creators, tested new apps weekly, and constantly refined his workflow. On paper, he was doing everything "right".

However, something felt off. Despite all the optimization, he was more exhausted than ever. He couldn't focus deeply. He jumped between tasks. He felt guilty whenever he wasn't "maximizing" his time.

The problem wasn't laziness. It was over-instrumented. Every part of his life had become a system to manage. AI hadn't freed his attention, it had fractured it.

When Tools Multiply, Attention Divides

There is a quiet lie at the center of modern productivity culture: "More tools equal more progress." In reality:

- Every tool requires setup.
- Every tool creates decisions.
- Every tool introduces maintenance.
- Every tool asks for attention.

AI tools are especially seductive because they promise leverage. However, leverage only works when it's focused.

When tools help you do one thing better, they serve you. When tools multiply options endlessly, they drain you. Clarity doesn't come from having more choices. It comes from having fewer, clearer ones.

Dopamine, Distraction, and the Modern Nervous System

To understand why technology exhausts us, we need to talk about dopamine, not as a buzzword, but as biology.

Dopamine is not the "pleasure chemical." It is the **anticipation chemical**.

It spikes when:

- You expect something new.
- You anticipate reward.
- You are uncertain about what comes next.

This is why:

- Notifications feel irresistible.
- Infinite scroll works.
- Constant novelty keeps you hooked.

AI-powered platforms are exceptionally good at this. They respond instantly. They adapt to you. They offer endless variation. Your brain does not experience this as neutral assistance. It experiences it as constant stimulation.

Burnout Isn't Always About Work

Many people think burnout comes from doing too much. Often, it comes from switching too much:

- Switching contexts.

- Switching tasks.
- Switching apps.
- Switching attention.

AI makes switching easier than ever.

You can:

- Generate an idea.
- Edit a draft.
- Start a new project.
- Abandon it.
- Start another.

All in minutes.

The problem is not speed. It is fragmentation. Your nervous system never gets to settle.

Real-Life Example: The Always-On Creative

Nina was a writer who embraced AI early. At first, it felt magical. Ideas flowed. Drafts appeared instantly. Possibilities expanded. However, over time, she noticed something unsettling.

She felt less connected to her work. She was producing more, but enjoying it less. Her mind was always "on," always scanning, always evaluating.

AI had removed friction, but also removed closure. Creativity became endless motion without depth.

The Cost of Infinite Possibility

Infinite possibility sounds liberating. In practice, it is exhausting. When everything is possible:

- Nothing feels finished.
- Nothing feels sufficient.
- Nothing feels worthy of rest.

AI does not naturally signal completion. Humans need endpoints.

We need moments where we can say: "This is enough for today." Without those boundaries, we drift into quiet burnout.

Why "More Tools" Doesn't Mean a Better Life

A better life is not the most optimized one. It is the most intentional one.

Technology excels at:

- Speed.
- Scale.
- Output.

Life thrives on:

- Meaning.
- Presence.
- Connection.

When tools become the center, life becomes a project to manage rather than an experience to inhabit.

AI should reduce complexity, not create new layers of it. If a tool adds mental overhead, it is not helping, no matter how powerful it is.

The Subtle Erosion of Attention

Attention erosion rarely announces itself. It looks like:

- Reaching for your phone without thinking.
- Opening an app and forgetting why.
- Feeling restless during quiet moments.
- Needing stimulation to feel okay.

These are not personal failures. They are predictable responses to environments designed to capture attention. AI amplifies this by making engagement smoother, faster, and more personalized.

The danger is not addiction. It is a dependency without awareness.

Real-Life Example: The Quiet Loss of Focus

Laura noticed she couldn't read anymore. She used to finish books easily. Now she rereads the same paragraph five times. She blamed herself. However, the issue was not intelligence or discipline.

Her attention had been trained for:

- Short bursts.
- Rapid feedback.
- Constant novelty.

Deep focus felt uncomfortable, not because she couldn't do it, but because she wasn't used to it anymore. AI had not caused the problem. However, it had normalized shallow engagement.

Mindful Use Starts With One Question

Before adopting any tool, especially AI, ask one question: "What problem in my life is this actually solving?"

I adopt a tool not because:

- It is impressive.
- It is popular.
- Everyone is using it.

But because:

- It does reduce stress.
- It does create clarity.
- It does give me back mental space.

If a tool doesn't help you, it doesn′t deserve your attention.

Setting Boundaries with Technology

Boundaries are not restrictions. They are protections. Healthy boundaries with AI and technology might include:

- No tools during rest time.
- No optimization of personal relationships.
- No constant tracking of personal habits.
- No replacing reflection with prompts.

Boundaries allow technology to stay in its lane.

Designing Intentional Tech Use

Intentional use looks different from person to person. For some, it means:

- Using AI only during work hours.

For others:

- Using AI only for specific tasks.

For families:

- No AI during meals or shared time.

The point is not rules. It is a choice.

Real-Life Example: Reclaiming Evenings

Marcus realized his evenings were disappearing. After work, he would "just check" a few tools, and suddenly it was midnight. He didn't delete everything.

He simply decided:

- AI tools off after 7 PM.
- Phone out of reach during dinner.
- No productivity content at night.

Within weeks, his sleep improved. His anxiety decreased. His sense of time returned. The tools hadn't changed. His relationship with them had.

Technology Should Expand Life - Not Replace It

AI can:

- Assist.
- Suggest.
- Organize.

- Clarify.

It cannot:

- Rest for you.
- Feel for you.
- Decide what matters.
- Live your life.

When technology begins to replace presence, something is off.

The Power of Boredom and Silence

Boredom is not a problem. It is a doorway. Silence is not empty. It is fertile. When AI fills every pause, we lose access to:

- Intuition.
- Creativity.
- Emotional processing.

Some of your best insights arrive when nothing is happening. Protect those moments.

You Are Allowed to Be Offline

You don't owe the world constant output. You don't owe tools constant engagement. You are allowed to:

- Rest without optimizing.
- Think without prompts.
- Create without assistance.
- Be without documenting.

AI should serve life, not consume it.

The Trap Is Not Technology - It is Forgetting Yourself

The greatest danger isn't AI controlling you. It is losing touch with your own internal signals. Your body knows when it is overwhelmed. Your mind knows when it is tired. Your attention knows when it is fragmented. Listening to those signals is a skill; ant technology cannot teach you this.

Choosing Presence over Performance

We live in an age that rewards performance. AI makes performance easier. However, presence is still human work.

Presence requires:

- Slowing down.
- Saying no.
- Choosing depth over speed.

AI should make presence easier, not harder.

The Real Question Going Forward

The future won't be decided by how powerful AI becomes. It will be decided by how consciously humans choose to live alongside with it.

The question is not:

"What can AI do?"

It is:

"What kind of life do I want to protect?"

You Are Still in Control - If You Choose to Be

Technology only controls you when you stop noticing. Attention is reclaimable. Boundaries are learnable. Presence is recoverable. AI doesn't need to disappear for that to happen. It just needs to stay in its place.

Pause for a Moment before Moving On

Attention shapes experience. What you give to others is what you live inside yourself.

Reflect:

- How do you usually feel after engaging with technology, energized or scattered?
- What habits quietly pull your attention away from what matters?
- Where could one boundary create noticeable calm?

Gentle Reminder:

Your attention is not infinite. Protecting it is not selfish, it is necessary.

9

ETHICS, TRUTH, AND RESPONSIBILITY (WITHOUT PHILOSOPHY HEADACHES)

Reflection Question:

If my name were attached to everything I shared, created, or trusted through AI, what would I slow down, double-check, or choose differently?

Ethics used to feel distant:

- Something debated in classrooms.
- Something handled by institutions.
- Something abstract and theoretical.
- Something abstract and theoretical.

However, AI has pulled ethics out of the lecture hall and placed it directly into everyday life.

Now ethics show up when:

- You forward an article you didn't fully read.
- You trust an AI-generated summary without checking.
- You post an image that looks real, but isn't.
- You let speed override accuracy.
- You choose convenience over care.

You don't need a philosophy degree to navigate this moment. You need attention, responsibility, and integrity.

This chapter is not about moral perfection. It is about staying awake in a world where truth is easier to blur than ever before.

Why Ethics Suddenly Feel Personal

Before AI, misinformation existed, but it traveled slower. Before AI, manipulation required effort. Before AI, most people didn't have the power to create convincing falsehoods at scale. That has changed.

Now:

- Anyone can generate realistic images.
- Anyone can simulate voices.
- Anyone can produce authoritative-sounding text.
- Anyone can flood the internet with plausible nonsense.

Ethics are no longer about what institutions do. Now ethics are about what you choose to create, share, believe, and amplify.

Deepfakes and Misinformation: Not a Future Problem

Deepfakes sound futuristic, but they are already here. They are not just celebrity videos or political scandals. They show up in:

- Fake job postings.
- Scam phone calls using familiar voices.
- Doctored screenshots.
- Altered images shared in-group chats.
- AI-generated news summaries missing context.

The danger isn't that people believe everything. It is that people stop knowing what to trust.

Real-Life Example: "I Thought It Was Real"

Karen received a video in a family group chat. It showed a public figure saying something shocking. The lighting looked real. The voice matched perfectly. The expressions felt human.

She felt angry. Betrayed. Alarmed. Before reacting, she paused and checked a reliable source. The video was fake.

She didn't feel relieved. She felt unsettled. Because if she could almost believe it, what about others? Deepfakes don't just spread lies. They erode confidence in reality itself.

Misinformation Doesn't Always Look Like Lies

One of the most dangerous assumptions is that misinformation is obvious. It isn't. Most misinformation today is:

- Partially true.

- Missing context.
- Emotionally charged.
- Framed to provoke reaction.

AI excels at producing content that sounds confident, neutral, and informed, even when it is wrong. That doesn't make AI evil. It makes human judgment essential.

Why Humans Remain Responsible

Here's the uncomfortable truth:

- AI doesn't remove responsibility. It redistributes it.
- AI doesn't decide what's shared.
- AI doesn't choose what goes viral.
- AI doesn't feel the consequences.

Humans do.

Every time you:

- Share content.
- Publish something.
- Use AI to persuade.
- Present AI output as fact.

You are making an ethical choice, whether you realize it or not.

"The AI Did It" Is Not an Excuse

Blaming AI is tempting:

- It feels external.
- It feels impersonal.
- It feels convenient.

However, responsibility doesn't disappear just because a tool is involved.

If you use AI to:

- Write a misleading email.
- Generate false claims.
- Create manipulative content.
- Spread unverified information.

The responsibility doesn't belong to the software. It belongs to the human who chose to use it that way.

Real-Life Example: The Shortcut That Backfired

Mark worked in marketing. Pressed for time, he used AI to draft promotional copy about a product's benefits. It sounded great, confident, polished, and persuasive. He didn't verify every claim.

Weeks later, customers complained. Some promises weren't accurate. Some expectations were inflated. Mark hadn't meant to deceive anyone. However, intention doesn't erase impact. AI had accelerated the message, but responsibility still landed on him.

Ethics Is Not About Perfection - It's About Care

Ethical behavior doesn't mean never making mistakes. It means:

- Checking before sharing.
- Pausing before amplifying.
- Correcting when wrong.
- Caring about consequences.

AI speeds things up. Ethics slow things down. That tension matters.

Teaching AI "Good Input" Starts With Us

AI systems learn from patterns. They reflect what we give them. That means:

- Biased prompts reinforce bias.
- Lazy prompts produce shallow output.
- Dishonest use encourages distortion.

AI doesn't invent values. It mirrors them.

Real-Life Example: Subtle Bias In, Subtle Bias Out

A hiring manager used AI to screen resumes. Without realizing it, the prompts emphasized "culture fit" and "strong leadership presence." The results skewed toward a narrow profile. Not because AI intended harm, but because the inputs carried assumptions.

When the prompts changed, focusing on skills and experience, the outputs changed too. Garbage in isn't just wrong information. Its unchecked assumptions.

Ethical Use Begins Before Output

Most people focus on the output. Ethical use starts earlier. It starts with:

- How you frame questions.
- What you ask for.
- What you prioritize.
- What you ignore.

AI reflects your intent more than your values. Align the two consciously.

Truth Requires Effort - Convenience Undermines It

Convenience is seductive. Why verify when you can summarize? Why read when you can skim? Why investigate when you can generate?

However, truth has always required effort. AI doesn't eliminate that requirement. It tests it. The easier it becomes to generate content, the more responsibility shifts to:

- Checking sources.

- Cross-referencing.
- Applying skepticism.
- Using discernment.

Real-Life Example: The Confidently Wrong Summary

Olivia relied on AI summaries to keep up with news. One day, she confidently discussed a topic, only to realize later that key context was missing. The summary wasn't lying. It was incomplete. However, incomplete information, presented confidently, is still misleading. Speed had replaced understanding.

Why "Neutral" Technology Isn't Neutral

AI is often described as neutral. However, neutrality is an illusion. AI reflects:

- Training data.
- Human choices.
- System incentives.
- Cultural norms.

It doesn't stand outside society. It is embedded within it. Which means ethics can't be outsourced.

Choosing Integrity Over Convenience

Integrity is not loud. It's quiet, often invisible, and rarely rewarded immediately. Integrity looks like:

- Saying "I don't know".
- Taking extra time to verify.
- Resisting viral outrage.
- Choosing accuracy over speed.

AI tempts us to move faster. Integrity asks us to slow down.

Real-Life Example: Not Posting First

During a breaking news event, Alex saw an AI-generated image circulating. It was dramatic. Emotional. Shareable. He almost reposted it. Instead, he waited. Hours later, the image was debunked. Alex didn't gain attention, but he kept his integrity. That matters more than metrics.

Ethics in Everyday AI Use

Ethics aren't only about deepfakes and global impact. They appear in small, daily decisions:

- Using AI to write a heartfelt message, and then sending it without reading.
- Letting AI speak for you when authenticity matters.
- Presenting AI-generated work as personal insight.
- Avoiding responsibility by blaming “the tool”.

These moments shape trust.

Trust Is Fragile in an AI World

Trust used to be built on:

- Familiar voices.
- Visual evidence.
- Authority.

AI disrupts all three.

Rebuilding trust requires:

- Transparency.
- Disclosure.
- Accountability.

You don’t need to announce every AI use. However, when it matters, be honest.

Teaching the Next Generation Responsibility

Children will grow up with AI as a given. What they need is not fear, but guidance. They need to learn:

- How to question outputs.
- How to verify information.
- How to value truth over convenience.
- How to use tools responsibly.

Ethics are learned through modeling, not lectures.

AI Doesn't Remove Morality - It Reveals It

AI exposes tendencies already present. If someone values speed over truth, AI amplifies that. If someone values integrity, AI supports that too.

The technology is powerful. However, the values behind it matter more.

The Real Risk Isn't Fake Content - It's Indifference

The most dangerous outcome isn't believing a lie. It is not caring whether something is true. When people stop caring about accuracy, ethics collapse quietly.

AI doesn't force indifference. However, AI can normalize it, if we let it.

You Are Still the Final Filter

No matter how advanced AI becomes:

- You choose what to believe.
- You choose what to share.
- You choose what to stand behind.

Responsibility doesn't scale away. It concentrates.

Ethics Without Headaches: One Simple Rule

Before using or sharing AI-generated content, ask: **"Would I stand behind this if my name were attached to it?"** If the answer is no, pause.

Integrity Is a Daily Practice

Ethics aren't grand gestures. They are daily habits. Like:

- Checking.
- Questioning.
- Slowing down.
- Caring.

AI makes ethics unavoidable, not optional.

Technology Changes - Values Endure

AI will evolve. Tools will improve. Capabilities will expand. However, the core question will remain: "What kind of person do I want to be in how I use this?" Answer that honestly, and let technology follow, not lead.

Pause for a Moment before Moving On.

Ethics aren't abstract anymore. They show up in everyday choices.

Reflect:

- When have you shared something without fully checking it?
- What helps you slow down before reacting or amplifying?
- What does integrity look like for you in small, ordinary moments?

Gentle Reminder:

You don't need to be perfect. You just need to care enough to pause.

10

HOW TO START USING AI WITHOUT OVERWHELM

Reflection Question:

What is one small, low-pressure area of my life where I could let curiosity replace fear, and what would "trying" look like if I didn't expect mastery?

If there is one emotion that comes up again and again when people talk about AI, it's not fear. It's overwhelming. Not the dramatic kind, the quiet kind.

The overwhelming that sounds like:

- "I don't even know where to begin."
- "There are too many tools."
- "I feel like I'm already behind."

- "What if I do it wrong?"

This chapter exists for that feeling. Not to push you forward faster, but to slow the whole thing down. Because the truth is this: You don't need to catch up. You don't need to master anything. You don't need to understand AI to start benefiting from it. You just need one gentle entry point.

The Lie That You Need to Learn Everything

One of the biggest barriers to starting with AI is the belief that it requires study. Need:

- Tutorials.
- Courses.
- Frameworks.
- Technical language.

That belief is wrong.

Most people who successfully use AI didn't "learn" it. They played with it. They treated it like a conversation, not a subject, and that makes all the difference.

One Simple Tool Is Enough

You don't need:

- Ten apps.
- Multiple subscriptions.
- A complex setup.
- A productivity system.

You need one tool that:

- You can access easily.
- You are not afraid to use.
- You can return to consistently.

That's it.

Starting with more than one tool almost guarantees overwhelm.

Real-Life Example: Too Many Options, No Progress

Rachel downloaded five AI tools in one weekend. She watched videos. She read comparisons. She created accounts everywhere. Two weeks later, she hadn't used any of them. Why?

Because every time she thought about starting, she had to decide *where* to start. Choice paralysis replaced curiosity. Eventually, she deleted all but one. Progress began, not because the tool was better, but because the decision was gone.

The Best First Tool Is the One You'll Actually Use

It doesn't matter which AI tool you choose first. What matters is:

- It feels approachable.
- It responds in plain language.
- You don't feel judged using it.

The goal of your first tool is not power. It is comfort. AI is not a race. It is a relationship.

Learning by Playing, Not Studying

Think about how you learned:

- To use a smartphone.
- To browse the internet.
- To send voice messages.

- To use social media.

You didn't read manuals.

You tapped. You tried. You made mistakes. You adjusted. AI works the same way. The fastest way to feel comfortable is not to learn about it, but to interact with it.

Low-stakes ways to play:

- Ask it to rewrite a sentence.
- Ask it to explain something simply.
- Ask it to brainstorm ideas.
- Ask it a question you already know the answer to.

Nothing breaks if it gets things wrong. That's the point.

Real-Life Example: From Intimidated to Curious

Carlos avoided AI for months. He felt like it was "for tech people." One evening, out of boredom, he asked: "Help me write a polite text declining an invitation." The response was… helpful. Not magical. Not life-changing. Just useful. The intimidation dissolved.

The next day, he asked another question. That's how it starts, not with ambition, but with curiosity.

Small Steps That Compound

AI doesn't change your life in one moment. It changes it through small, repeated wins. A clearer email. A faster plan. A better explanation. A moment of relief.

These don't feel dramatic, but over time, they compound into:

- Less mental friction.
- More confidence.
- Greater ease with technology.

Consistency matters more than intensity.

Think "Assist," Not "Automate"

One mistake beginners make is trying to automate everything immediately. That leads to complexity, confusion, and detachment. A better mindset is assistance.

Let AI:

- Help you start.
- Help you clarify.
- Help you organize.

However, keep yourself in the loop. The goal is support, not replacement.

Real-Life Example: The Over-Automation Trap

Linda wanted AI to handle her entire workflow. She built prompts. She created systems. She followed optimization guides. Soon, she felt disconnected from her own work. She didn't understand the output anymore.

So she reset. She used AI only to:

- Draft first versions.
- Organize thoughts.
- Reduce friction.

Her work felt human again.

When Not to Use AI

This matters as much as when to use it. AI is not always the right tool. You may want to avoid AI when:

- Processing strong emotions.

- Making deeply personal decisions.
- Navigating relationships.
- Sitting with uncertainty.
- Resting or reflecting.

AI is excellent at doing. It isn´t meant for being. Silence, boredom, and reflection still matter.

AI Should Make Things Lighter - Not Heavier

A simple test. After using AI, ask yourself: “Do I feel clearer or more cluttered?”

If you feel more anxious, pressured, or overwhelmed, that is a signal. You are not failing. You are using it in a way that doesn’t serve you. Adjust.

Permission to Be Bad at It

You will:

- Ask unclear questions.
- Get irrelevant answers.
- Feel awkward at first.
- Wonder if you are “doing it right”.

Good. That means you are learning. AI doesn't reward perfection. It responds to experimentation.

Real-Life Example: Asking the "Wrong" Way

Tom kept getting unhelpful answers. He assumed AI wasn't useful. Later, he realized his questions were vague. Instead of: "Help me with this," he tried: "Help me rewrite this email to sound calm and professional." The difference was immediate. The tool hadn't changed. The interaction had.

Start Where the Friction Is

The best place to begin is where you already feel resistance. Ask yourself:

- What task do I procrastinate most?
- What drains my energy?
- What feels heavier than it should?

Start there. AI doesn't need to touch every part of your life. Just one.

Real-Life Example: One Pain Point Is Enough

Jenna hated writing emails. Not all writing, just emails. She started using AI only for that. Nothing else. Within weeks, she felt more confident at work. She didn't become an "AI user." She solved a problem.

You Are Not Late

This matters enough to repeat. You are not behind. Most people are still figuring this out. Starting slowly is not a weakness. It is wisdom.

Pause for a Moment Before Moving On

Beginning doesn't require confidence. It requires willingness.

Reflect:

- Where does curiosity feel safer than ambition?
- What's one low-stakes area of your life where you could experiment gently?
- What would "trying" look like without expecting mastery?

Gentle Reminder:

You are allowed to start small. Progress compounds quietly.

11

STAYING RELEVANT IN AN AI WORLD

Reflection Question:

Instead of asking how I can keep up, what would change if I asked how I want to keep contributing, and which of my human strengths am I underestimating?

Relevance used to feel stable. You learned a profession. You earned credentials. You gained experience. You climbed gradually. For a long time, that was enough.

AI has unsettled something deeper than job titles or tools. It has unsettled the idea that relevance is something you achieve once and then keep.

Now relevance feels fragile. People ask:

- "Will my skills still matter?"
- "What if AI replaces what I do?"
- "Am I too late to adapt?"
- "What should I even learn now?"

This chapter isn't about predicting the future. It is about understanding what doesn't go obsolete, and why relevance has always been less about what you know and more about how you move through change.

The Fear Behind the Question of Relevance

When people say they are afraid of becoming irrelevant, they are rarely talking about technology.

They're talking about:

- Losing dignity.
- Losing income.
- Losing purpose.
- Losing a sense of contribution.

Work is not just work. It is identity, stability, and meaning. AI doesn't threaten human worth, but it does threaten static identities. And that is uncomfortable.

If you have built your life around being competent, useful, and needed, the idea of becoming "replaceable" feels deeply personal.

The goal here isn't to deny that fear. It is to put it in the right place, so it doesn't drive your decisions.

What AI Changes - And What It Doesn't

AI changes:

- Speed.
- Access to information.
- How tasks are executed.
- Which tasks are automated.

AI does **not** change:

- Human judgment.
- Responsibility.
- Meaning-making.

- Trust.
- Context.
- Ethics.
- Care.

The mistake many people make is assuming relevance lives in tasks. It doesn't. Tasks change constantly. Relevance lives in capabilities, the human strengths that remain valuable even as tools evolve.

The New Divide Isn't "Tech" vs. "Non-Tech"

The future won't be split between "AI people" and "non-AI people."

It will be split between:

- People who stay rigid.
- People who stay responsive.

Responsiveness doesn't mean chasing every trend. It means staying mentally flexible, curious, and willing to update how you work. That is not a technical skill. It's a life skill.

Skills That Will Matter More, Not Less

Ironically, the more capable AI becomes, the more certain human skills increase in value. Not technical skills alone, human ones.

1. Judgment

AI can generate options. It cannot decide what matters.

Judgment is the ability to:

- Weigh trade-offs.
- Understand context.
- Recognize nuance.
- Choose wisely under uncertainty.

Judgment develops through experience, reflection, and responsibility, not automation.

Real-Life Example: The Manager Who Became More Valuable

Sophie managed a team in a data-heavy industry. When AI tools began generating reports instantly, she felt the ground shift beneath her. If the data could be produced in minutes, what was she for?

Then she realized something. Her team didn't need more information. They needed:

- Interpretation.
- Prioritization.
- Decisions.

Sophie leaned into the questions AI couldn't answer:

- "What matters right now?"
- "What's the risk of acting too fast?"
- "What are the second-order consequences?"
- "What would we regret doing?"

AI provided information. She provided judgment. And in the end, the person who decides wisely remains indispensable.

2. Communication

AI can write, but communication is not writing. Communication is:

- Understanding your audience.
- Reading emotional cues.
- Explaining complexity simply.

- Navigating disagreement.
- Building trust.

AI can assist with words. It cannot own relationships.

Real-Life Example: The Engineer Who Learned to Translate

David was technically brilliant but struggled to explain his ideas. For years, that didn't matter. His expertise carried him.

Then AI entered his workplace, and technical output became easier for everyone. His advantage shrank. So he adapted.

He used AI to help him:

- Simplify explanations.
- Create clearer slides.
- Practice presentations.
- Translate complex ideas into plain language.

And he did the human part:

- Reading the room.
- Adjusting tone.

- Handling objections.
- Building credibility.

His relevance didn't come from knowing more. It came from helping others understand.

3. Adaptability

Adaptability isn't about being comfortable with change. It's about being willing to learn while uncomfortable.

Adaptable people:

- Experiment without mastery.
- Ask questions without shame.
- Adjust without clinging.
- Stay curious under pressure.

AI rewards adaptability because it evolves constantly. Those waiting for stability will struggle. Those willing to move will not. Adaptability is not personality. It is practice.

Why Technical Skills Alone Are Not Enough

Learning AI tools is useful, but insufficient. Tools change quickly.

If your relevance depends on:

- One platform.
- One system.
- One workflow.

You are vulnerable.

However, if your relevance depends on:

- Thinking clearly.
- Communicating well.
- Learning continuously.

You are resilient. A tool can become outdated. A mindset doesn't.

Learning How to Learn Again

One of the most important skills in an AI world is meta-learning: Learning how to learn. Many adults haven't learned something new deeply in years.

They are used to:

- Mastery.
- Expertise.
- Competence.

AI reintroduces beginnerhood. That is humbling and necessary.

Real-Life Example: Starting From Zero at 45

Angela had been successful in her career for decades. When AI tools entered her field, she felt exposed. She hated not knowing. She felt behind.

Her instinct was to protect her identity as “the expert.” Instead, she chose curiosity. She asked younger colleagues questions. She experimented privately. She allowed herself to be slow.

Within a year, she wasn’t an expert, but she was fluent. Her relevance didn’t come from speed. It came from willingness.

Beginner's Mindset Is Not a Weakness

In many professional cultures, being a beginner feels risky. In an AI world, it is a strength.

Being a beginner allows you to:

- Learn continuously.
- Adapt quickly.
- Avoid rigid thinking.
- Stay open to new roles.

The most irrelevant people are not the least skilled. They are the most rigid.

Relevance Is a Mindset, Not a Degree

Degrees represent past learning. Relevance requires present engagement. No credential guarantees relevance. No age disqualifies you from it.

Relevance is not about:

- Knowing everything.
- Chasing trends.

- Becoming technical overnight.

It's about:

- Staying mentally flexible.
- Updating assumptions.
- Remaining curious.
- Taking responsibility for growth.

Why Experience Still Matters

AI has access to information. It does not have experience. Experience includes:

- Knowing what fails quietly.
- Understanding human reactions.
- Recognizing patterns over time.
- Anticipating consequences.

These cannot be downloaded.

Experienced people who adapt become anchors in change, not because they know more facts, but because they know what facts mean when humans enter the equation.

The Danger of Outsourcing Thinking

One quiet risk in an AI world is mental atrophy. When people rely on AI to:

- Decide.
- Judge.
- Prioritize.
- Interpret.

They lose sharpness.

Use AI to assist thinking, not replace it. If AI becomes your default brain, your brain will slowly stop leading.

Relevance Requires Responsibility

Staying relevant today also means staying ethical, aware, and accountable.

People who:

- Question outputs.
- Care about truth.
- Consider consequences.

They will always matter, because responsibility cannot be automated.

Real-Life Example: The Trusted Voice

When AI-generated content flooded a company's internal systems, confusion grew.

Lona became known for something simple:

- She checked facts.
- She clarified context.
- She asked thoughtful questions.

She wasn't the fastest. She was the most trusted. Trust is relevant.

The Myth of "Keeping Up"

You don't need to keep up with everything. That's impossible.

You need to:

- Stay oriented.
- Understand fundamentals.
- Know where to look when needed.

Relevance doesn't mean speed. It means direction.

Choosing Depth Over Novelty

AI creates constant novelty. Relevance comes from depth.

People who:

- Go deep in understanding.
- Build long-term insight.
- Connect ideas across domains.

They remain valuable even as tools change. Novelty makes noise. Depth makes sense.

You Are Not Competing With AI

You are collaborating with it. And in that collaboration:

- Your judgment matters
- Your voice matters
- Your integrity matters

The future belongs to people who stay human on purpose.

Relevance Is About Contribution

The deepest question is not: "How do I avoid being replaced?" It is: **"How do I continue to contribute meaningfully?"** Answer that and relevance follows.

Pause for a Moment Before Moving On

Relevance is not a race. It's a relationship with change.

Reflect:

- Which of your strengths feel increasingly important, not obsolete?
- How do you respond when you don't know something yet?
- What would learning look like if you didn't judge yourself for being a beginner?

Gentle Reminder:

Relevance doesn't come from knowing everything. It comes from staying open.

12

A HUMAN FUTURE WITH AI

Reflection Question:

If technology keeps getting faster, what parts of my life do I want to protect from speed, and what kind of human do I want to be while everything else accelerates?

Every generation believes it stands at the edge of something unprecedented. Sometimes that belief is exaggerated. This time, it isn't.

AI is not just another tool layered onto life. It is a force that reshapes how we think, work, create, decide, and even how we understand ourselves.

That's why the most important conversation about AI isn't technical. It is human.

The question we face is not whether AI will advance. It will. The real question is whether we will remain present, grounded, and intentional as it does.

AI as a Mirror of Humanity

AI does not arrive with values. It arrives with patterns. It reflects what we feed it:

- Our language.
- Our priorities.
- Our biases.
- Our incentives.
- Our impatience.
- Our brilliance.

When people say AI is dangerous, what they often mean is that it reveals things about us we would rather not see. It mirrors our shortcuts. Our obsession with speed. Our discomfort with uncertainty. Our hunger for validation.

AI is not a replacement for humanity. It is a reflection of it amplified.

What AI Shows Us About Ourselves

When AI produces shallow content, it's because we rewarded shallow content. When AI spreads misinformation, it's because truth was already undervalued. When AI optimizes everything, it's because we equated efficiency with worth long before machines entered the room.

This is uncomfortable and powerful. Because mirrors don't just expose flaws. They offer a chance to adjust.

Real-Life Example: The Automation Wake-Up Call

Ellen worked at a fast-growing company that embraced AI enthusiastically. Emails were automated. Reports were generated instantly. Meetings were summarized. Decisions were accelerated.

On paper, productivity soared. However, something else happened. People stopped asking questions. Nuance disappeared. Careful thinking was replaced by fast output.

Then a major decision technically correct and efficiently executed, triggered a public backlash. The data had been accurate. The timing had been optimized. However, the human impact hadn't been considered.

AI hadn't failed. Human judgment had been sidelined. That moment didn't lead the company to reject AI. It led them to restructure how they used it, keeping speed, but re-centering responsibility.

The Danger of Outsourcing Meaning

AI is exceptionally good at handling *how*. It can tell you:

- How to optimize.
- How to scale.
- How to grow.
- How to improve.

However, meaning lives in *why*. Why does this matter? Why is this worth doing? Why does this align with who you are? Outsourcing meaning is tempting because meaning is hard. Meaning requires reflection.

Values. Uncertainty. Patience. AI offers answers. Meaning requires questions.

When Life Becomes a System to Manage

There is a quiet risk in letting AI organize every corner of life. When everything is optimized:

- Meals.
- Schedules.
- Workflows.
- Goals.
- Habits.

Life can slowly turn into a dashboard.

You are productive. You are efficient. You are on track. However, you are not necessarily fulfilled, because meaning doesn't always show up in metrics. Some of the most important moments in life are inefficient by design.

Real-Life Example: The Perfectly Optimized Burnout

Jason was the kind of person AI tools were made for. He tracked everything. Optimized routines. Used AI for planning, writing, thinking, and decision-making.

His days ran smoothly. Too smoothly. One evening, he realized he couldn't remember the last time he felt surprised, moved, or deeply engaged.

Nothing was wrong. Nothing was alive either. He hadn't lost control. He had lost contact with his own experience. Jason didn't abandon AI. He stopped letting it design his life. He kept the tools, but reclaimed the steering wheel.

Choosing Presence Over Speed

Speed is seductive. AI makes everything faster:

- Writing.
- Planning.
- Deciding.
- Responding.

However, speed has a cost.

Presence requires slowness. Presence means:

- Listening fully.
- Thinking deeply.
- Feeling without distraction.
- Responding intentionally.

You cannot rush presence. No technology can generate it for you.

Why Faster Isn't Always Better

Many of life's most meaningful experiences resist acceleration:

- Grief.
- Love.
- Learning.
- Healing.
- Creativity.
- Trust.

AI can support the logistics around these experiences. It cannot replace the experiences themselves.

When we chase speed everywhere, we risk skipping the very things that make life worth living.

Real-Life Example: The Conversation That Couldn't Be Optimized

Maria used AI to help her with almost everything, except one thing. When her relationship began to struggle, she was tempted to:

- Draft messages.
- Analyze conversations.
- Optimize responses.

Instead, she stopped.

She chose presence. She listened. She stumbled. She spoke imperfectly. The conversation was slow, awkward, and emotional. It was also real. Some moments demand humanity, not efficiency.

Designing a Life, Not Just Efficiency

Efficiency answers the question: "How can I do more with less?" Design answers a deeper one: "What kind of life do I want to live?"

AI is excellent at efficiency. Life design requires values. Designing a life means deciding:

- What deserves attention.
- What can be ignored.
- What matters enough to protect.
- What pace feels sustainable.
- What success actually means to you.

No algorithm can answer these for you.

Real-Life Example: Redefining Success

Lena had always defined success as output. More work. More growth. More achievements. AI amplified that trajectory. She could do more than ever, but something felt off.

Therefore, she redesigned her definition of success. Success became:

- Time with family.
- Creative depth.
- Fewer commitments done well.
- Space to think.

AI didn't disappear. It was adjusted. Technology followed values, not the other way around.

AI Is a Tool - But Life Is Not a Project

Projects end. Life doesn't. When life becomes a project:

- There is constant evaluation.
- Endless improvement.
- No arrival point.

This creates quiet dissatisfaction.

Designing a life means allowing it to be lived, not endlessly optimized. AI should support life's rhythms, not override them.

The Role of Choice in a Technological World

AI expands options. Choice becomes heavier. Without intentional values, more options create more anxiety, not freedom.

Choosing intentionally means:

- Saying *no* to unnecessary tools.

- Limiting optimization.
- Embracing constraints.
- Protecting unproductive time.

Freedom is not infinite choice. It is a meaningful choice.

A Human Future Is Not Automatic

A human future with AI is not guaranteed. It's a choice made daily, quietly, and often invisibly. It shows up when:

- You pause instead of react.
- You verify instead of share.
- You reflect instead of optimize.
- You stay present instead of productive.

These choices don't trend, but they shape culture.

AI Will Keep Getting Better at Being Machine

AI will:

- Get faster.
- Get smarter.
- Get more integrated.

That's inevitable.

What's not inevitable is whether humans:

- Stay reflective.
- Stay responsible.
- Stay connected.
- Stay grounded.

That part is up to us.

The Quiet Power of Being Human on Purpose

Being human used to be the default. Now it's practice. It means:

- Thinking before delegating.
- Feeling before optimizing.
- Choosing meaning over metrics.
- Designing boundaries with intention.

AI doesn't remove humanity. It challenges us to claim it deliberately.

Reclaiming Energy Means Respecting Human Limits

AI does not get tired. You do.

AI does not need rest. You do.

One of the quiet dangers of powerful tools is that they raise expectations without regard for human limits. More speed. More availability. More responsiveness. More productivity. However, energy is not infinite.

Using AI wisely means:

- Letting it carry what drains you.
- Protecting what energizes you.
- Recognizing when "more" becomes too much.

AI should reduce exhaustion, not normalize it.

Focus Is the New Luxury and the New Responsibility

In a world of infinite inputs, focus is no longer automatic. It must be protected.

AI can either:

- Multiply distractions.
- Or remove noise.

The difference lies in how you use it.

Focus grows when:

- Tools are limited.
- Notifications are intentional.
- Boundaries are respected.
- Silence is allowed.

AI should help you think more clearly, not keep you constantly stimulated.

Using AI Intentionally Is a Daily Practice

There is no final setup. No perfect system. No moment where you "figure it out" forever. Intentional use is ongoing.

It looks like asking simple questions again and again:

- "Is this helping or distracting me?"
- "Is this saving energy or draining it?"
- "Is this aligned with how I want to live?"

Sometimes the answer will be yes. Sometimes it won't. Wisdom is adjusting, not doubling down.

You Don't Need to Keep Up - You Need to Choose Wisely

The world will continue to shout:

- New tools.
- New updates.
- New trends.
- New fears.

You are not obligated to chase any of it.

Relevance is not about speed. It's about alignment. You don't need to know everything. You don't need to use everything. You don't need to understand everything. You need to choose what supports *your* life, not someone else's vision of success.

Comparison Is the Fastest Way to Lose Yourself

One of the hidden costs of an AI-driven world is comparison.

Someone is always:

- Producing more.
- Scaling faster.
- Automating better.
- Adapting sooner.

However, comparison ignores context. Your life is not a benchmark. Your pace is not a flaw. Your priorities are not outdated. Technology should adapt to humans, not the other way around.

Technology Should Serve Life, Not Replace It

This is the core truth beneath everything in this book. Technology is a means. Life is the end.

AI should:

- Support relationships, not replace them
- Free mental space, not fill it
- Enhance creativity, not drain it
- Assist judgment, not override it

When technology starts defining your worth, pace, or identity something has gone wrong. Not because technology is evil, but because tools were never meant to lead.

The Quiet Power of Choosing Enough

AI makes “more” always possible. More output. More ideas. More options. More efficiency. However, fulfillment often comes from choosing *enough.*

Enough work. Enough growth. Enough information. Enough productivity. Enough is not settling. It is deciding.

You Are Still the One Living the Life

No matter how advanced AI becomes:

- It doesn’t feel your days.
- It doesn’t carry your values.
- It doesn’t experience love, loss, or meaning.
- It doesn’t define a good life.

You do.

AI can support you, but it cannot replace your role as the author of your life.

The Future Is Not Something That Happens to You

The future does not happen to you. Millions of small, ordinary decisions shape your future. What you automate. What you protect. What you slow down. What you refuse to optimize. What you choose to be present for. None of those decisions makes headlines. All of them matter.

A FINAL REFRAMING

The question is not: "How do I stay ahead of AI?" It's: "How do I stay human while using it?" The answer is not technical. It is intentional.

You Are Allowed to Design a Slower, Deeper Life

Despite what the world suggests:

- You don't need to be constantly available.
- You don't need to maximize every hour.
- You don't need to optimize everything you touch.

You are allowed to:

- Think slowly.
- Choose depth.
- Rest without justification.

- Create without pressure.

AI does not revoke that permission. Only forgetting yourself does.

This Is Not the End - It's a Beginning

You don't finish this book with a system. You finish it with clarity.

Clarity about:

- What AI is good at.
- What humans are irreplaceable at.
- What kind of life you want to protect.

That clarity is enough.

One Last Thought

Smarter machines will keep coming. That part is inevitable. Better lives are not.

A better life require:

- Choice.
- Boundaries.
- Reflection.

- Courage.

You don't need to control technology. You need to relate to it wisely.

The Goal Was Never Smarter Machines

The goal has always been:

- More time for what matters.
- More energy for living.
- More focus for meaning.
- More presence for being human.

If AI helps with that, use it. If it doesn't, step back. You are not here to serve technology. Technology is here to serve life. And your life is worth designing with care.

WHAT COMES NEXT

AI Without Fear is the first book in a three-part exploration of how we understand, use, and live alongside *artificial* intelligence (AI).

The books that follow move gently forward, first into practical daily use, and later into protecting what gives life meaning in an intelligent age.

This journey isn't about keeping up. It is about choosing wisely.

Living Well with AI - A Trilogy

AI Without Fear

Using AI Well

Staying Human

Made in the USA
Monee, IL
04 November 2026

67382705R10089